BEGINNING HISTORY

EGYPTIAN FARMERS

Jim Kerr

Illustrated by Bernard Long

Wayland

BEGINNING HISTORY

The Age of Exploration
The American West
Crusaders
Egyptian Farmers
Egyptian Pyramids
Family Life in World War II
Greek Cities
The Gunpowder Plot
Medieval Markets
Norman Castles

Plague and Fire
Roman Cities
Roman Soldiers
Saxon Villages
Tudor Sailors
Tudor Towns
Victorian Children
Victorian Factory Workers
Viking Explorers
Viking Warriors

All words that appear in **bold** are explained in the glossary on page 22.

Series Editor: Rosemary Ashley
Designer: Helen White

First published in 1990 by Wayland (Publishers) Limited, 61 Western Road,
Hove, East Sussex BN3 1JD

© Copyright 1990 Wayland (Publishers) Limited

2nd impression 1991

British Library Cataloguing in Publication Data
Kerr, Jim
Egyptian farmers.
1. Egypt. Agriculture ancient period
I. Title   II. Series
630′.932

ISBN 1-85210-906-8

Typeset by Kalligraphics Limited, Horley, Surrey.
Printed in Italy by G. Canale & C.S.p.A., Turin.
Bound in Belgium by Casterman S.A.

CONTENTS

ANCIENT EGYPT

Egypt is a hot, dry country. Most of
the land is desert and not suitable for
growing crops.

The Ancient Egyptians lived on
two narrow strips of land on either

side of the River Nile. They settled there because once a year the river flooded, soaking the land and providing very rich, **fertile** soil for their crops.

The people of Ancient Egypt were ruled by a King, called a Pharaoh, who owned all of the land.

FARMING IN ANCIENT EGYPT

Egyptians learnt how to grow crops and keep herds of animals about 10,000 years ago. These early farmers lived near the Nile, where the soil was fertile.

Below *The banks of the River Nile.*

The farmers grew fruit such as apples, dates, figs, grapes and olives. They also grew vegetables, wheat, barley and flax which was used for making linen.

Egyptian farmers divided the year into three seasons: the flooding season, the ploughing and planting season and the harvesting season. At the end of the third season, farmers had to pay a **tax** to the Pharaoh.

Above *Egyptian farmers using reed boats to transport produce along the Nile.*

7

THE ANNUAL FLOOD

Every summer, heavy rains fell on the land south of Egypt. The rainwater poured into the River Nile which flowed north to Egypt and flooded the land. **Nilometers** measured how high the Nile's water had risen.

When the flood-water drained away, it left behind a layer of damp **silt** that was ideal for growing crops.

Sometimes, a farmer would complain that his crops did not grow because the floods had not reached his land. This meant he could not pay his taxes. The tax collector would check a Nilometer to see if the farmer was telling the truth.

PREPARING THE LAND

After the floods fell, everybody helped
to prepare the land for planting.
First, men, women and children
picked up all the stones and dead
branches left behind by the floods.
Next, they built low mud-walls to
mark the boundaries between the
different fields.

Above *A wooden model of an Egyptian farmer ploughing the land.*

Below *A farmer hoeing the soil.*

The soft soil was easy to **till**. To turn the soil over, farmers used **hoes** and ploughs drawn by oxen. One person walked ahead of the oxen, scattering seeds. The oxen trod the seeds into the ground and the plough covered the seeds with soil.

CANALS AND DITCHES

Short canals were dug from the river to the fields. When the river flooded, water ran along these canals and filled storage ditches. After the floods fell, farmers slowly let the water out of the ditches to **irrigate** their fields of crops.

Below *A shaduf.*

Farmers watered their crops by

means of shadufs. These were swinging beams with a heavy weight on one end and a bucket on the other. Farmers pulled the buckets down and dipped them in the river. The weights helped farmers to lift the buckets.

Above *This painting of irrigation canals comes from the walls of an Egyptian tomb.*

HARVESTING CROPS

Egyptian farmers harvested their wheat and barley crops in March or April. The crops had to be cut, gathered, **threshed** and separated.

They cut the stalks with **sickles**, then gathered up the ears of grain.

These were spread on the ground and threshed, either by beating them with **flails** or by driving oxen and sheep over them. This produced a mixture of grain and **chaff**.

They separated the mixture by tossing it into the air, using small scoops. The wind blew the light chaff away, leaving the grain behind.

They picked the fruit from trees

15

ANIMAL FARMING AND HUNTING

Below *Catching wild birds in the marshes.*

Farmers raised herds of antelopes, cattle, goats, pigs and sheep. They also bred birds – such as ducks and geese – which they fattened up by forcing food down their throats.

Oxen and donkeys were used for pulling and carrying heavy loads. Dogs guarded farm animals and were trained to help hunters catch wild birds and animals. Cats were kept for catching mice and rats.

Rich people used to hunt wild animals in the desert. They also used reed boats to fish and hunt for wild birds in the marshes.

Below *In this wall painting, you can see Egyptian farmers herding and checking cattle.*

WHAT PEOPLE ATE

Most farming families ate fairly simple food. A poor farmer would eat only bread, cheese, beans and salad, and drink water.

Wealthier people had more varied diets. They could choose from the many kinds of bread and cakes produced by bakers. They also ate beef, pork, antelope and hyena meat, as well as goose, pigeon and fish from the Nile. They drank beer or wine.

Adults ate their food sitting on simple stools; children sat on the floor. Everybody ate with knives and their fingers.

Below *This wooden model is about 4000 years old. You can see a butcher, a baker and a brewer.*

Left *Can you see how these butchers have tied up the cow's legs before they kill it?*

19

HOW THEY LIVED

Farm workers lived in simple houses. Few of their houses have survived because they were built of mud, which has long since crumbled away.

A typical worker's house had a hall, a living room, a kitchen and a bedroom. The windows were small, to keep out the hot sun. Stairs led to a flat roof where there was a shed with matting walls. This made an extra room on top of the house.

People wore few clothes because it was so warm. Men wore kilts, women wore ankle-length dresses and small children went naked.

GLOSSARY

Chaff The unwanted parts of plants (husks etc.) after harvesting.

Fertile Able to feed plants.

Flails Tools, like whips.

Hoes Tools used to till the soil.

Irrigate To supply land with water.

Nilometers Instruments used to measure the height of the River Nile.

Sickles Tools with curved blades, used for cutting grass.

Silt Thick black mud.

Tax Money or produce which is given to the rulers of a country.

Thresh To remove grain from unwanted parts of plants after harvesting.

Till To prepare land for growing crops.

BOOKS TO READ

A Pocket Guide to Ancient Egypt by Anne Millard (Usborne, 1981).

Cleopatra and the Egyptians by Andrew Langley (Wayland, 1985).

The Egyptians by Lucilla Watson (Wayland, 1986).

INDEX

Picture acknowledgements

The publishers would like to thank the following for providing the photographs in this book: Bruce Coleman 6 (bottom); Michael Holford 11 (top), 16 (bottom), 17 (bottom), 18 (bottom), 19 (top); Ronald Sheridan 11 (bottom), 13 (top).